# YOUR KNOWLEDGE HAS VALUE

- We will publish your bachelor's and master's thesis, essays and papers

- Your own eBook and book - sold worldwide in all relevant shops

- Earn money with each sale

Upload your text at www.GRIN.com and publish for free

**Bibliographic information published by the German National Library:**

The German National Library lists this publication in the National Bibliography; detailed bibliographic data are available on the Internet at http://dnb.dnb.de .



**Imprint:**

Copyright © 2015 GRIN Verlag, Open Publishing GmbH
Print and binding: Books on Demand GmbH, Norderstedt Germany
ISBN: 9783668494169

**This book at GRIN:**

http://www.grin.com/en/e-book/371529/conjugate-gradient-method-for-the-solution-of-optimal-control-problems

Henry Ekah-Kunde

# Conjugate gradient method for the solution of optimal control problems governed by weakly singular Volterra integral equations with the use of the collocation method

GRIN Publishing

# Conjugate gradient method for the solution of optimal control problems governed by weakly singular Volterra integral equations with the use of the collocation method

by
Henry E. Ekah-Kunde
SAINT MONICA UNIVERSITY BUEA

November 2015

## Abstract

*In this research, a novel method to approximate the solution of optimal control problems governed by Volterra integral equations of weakly singular types is proposed. The method introduced here is the conjugate gradient method with a discretization of the problem based on the collocation approach on graded mesh points for non linear Volterra integral equations with singular kernels. Necessary and sufficient optimality conditions for optimal control problems are also discussed. Some examples are presented to demonstrate the efficiency of the method.*

**Keywords:** Optimal control problem; Volterra integral equation; Collocation method; Conjugate gradient method; Numerical method.

## 1 Introduction

Optimal control problems appear in a wide range of applications in engineering and science. Many problems in epidemiology, biology, economics and the like belong to the class of optimal control problems governed by (Volterra) integral equations.

In the last couple of decades advances in the solution of optimal control problems have been made. Different types of methods have been proposed to solve optimal control problems of this class.

A general model of an optimal control problem of this type involves the minimization of an objective function governed by an integral equation, which contains and depends on control functions that are selected within certain limits. The objective here is to solve for the control that satisfies the state equation and its restrictions, and with the aid of this control as well as the desired state, the solution of the integral equation, minimize the objective function.

Our control is one that is governed by Volterra integral equation with weakly singular kernel. A treatment of the solution of Volterra integral equations has been handled in several articles, for instance in the works of Brunner [7], Brunner and van der Houwen [11] and Te Riele [22].

A theoretical analysis of the optimal control problem involves the investigation of the existence of a solution to it, that is, the treatment of necessary and sufficient optimality conditions. In the handling of these conditions problems will be encountered with respect to the basic space. The integral operators occurring in this problem are differentiable only in the $L_\infty$ space, while the sufficient optimality conditions can only be satisfied in the $L_p$ (generally $p = 2$) spaces. To overcome this obstacle we shall apply the "Two Norm Technique", which has been perfectly handled in the works of Maurer [19] and Tröltzsch [23] and [24]. We shall discuss briefly on the necessary optimality conditions as well as the regularity conditions, which are prerequisites for optimality and the Lagrange multiplier rule. Sufficient optimality conditions have been treated in several articles such as the publications of Maurer [19], Maurer and Zowe [20], Goldberg and Tröltzsch [15], Casa, Tröltzsch and Unger [12] and Tröltzsch [23], [24] and [25]. The approach used by Casas, Tröltzsch and Unger [12] for nonlinear elliptical boundary control problems shall be applied to our problem. These theoretical investigations provide a basis for a numerical treatment of the optimal control problems, which is the subject matter of this work. The numerical method chosen for solving optimal control problems here is the conjugate gradient method.

# 2 Optimality conditions for optimal control problems governed by a finite number of integral equations

## 2.1 A class of optimal control problems

We shall first start by describing a class of optimal control problems that will be considered in a greater part of this work:

$$\text{(OCP)} \qquad \text{Minimize} \quad \int_0^T f(x(t), u(t), t)$$

$$\text{subject to} \quad x(t) = c(t) + \int_0^t k(t, s) b(x(s), u(s), s) ds \tag{1}$$

$$\int_0^T g_i(x(s), u(s), s) ds - c_i = 0, \qquad i = 1, \cdots, k,$$

$$u_a(t) \leq u(t) \leq u_b(t) \qquad a.e. \quad \text{on} \quad [0, T],$$

with the given functions defined as follows:
$f, g_i, b : R \times R \times [0, T] \to R$, $c : [0, T] \to R$, whereby $u_a(t)$ and $u_b(t)$ are respectively the lower and upper limits of our control function, with $-\infty < u_a$ and $\infty > u_b$, and $T > 0$ is fixed given value.

Furthermore, we make the following assumptions with respect to the functions featuring in the control problem (1):

## [A1]

The kernel, $k(t, s)$ is continuous on $\{(t, s) \in [0, T] \times [0, T] \mid 0 \leq s \leq t \leq T\}$, and at $t = s$ it is bounded by
$$k(t, s) \leq c(t - s)^{-\alpha},$$

with $c \in R$ and $\alpha \in (0, 1)$. Hence $k(t, s)$ is weakly singular at $t = s$.

## [A2]

The functions $f, g_i, b$ defined on $R \times R \times [0, T]$ are twice partially differentiable with respect to $x$ and $u$ and their derivatives are continuous on $R \times R \times [0, T]$ with respect to $x$ and $u$.

## [A3]

Furthermore, these functions and their derivatives up to the second order are Lipschitz continuous with respect to $x$ and $u$ on $R \times R \times [0, T]$.
In equation 1, $u(t)$ is known as the control function with $u \in L_\infty(0, T)$ and

$x(t)$ is said to be the state corresponding $u(t)$ with $c(.) \in C[0,T]$.

<u>**Comment**</u> : [A2] can be weakened to well-known Caratheodory type conditions. Moreover, [A3], which requires global Lipschitz continuity, can be weakened to an associated local assumption. We avoid this for the sake of easier presentation.

The optimal control problem (OCP) described in (1) can also be represented abstractly in operator form as follows:

$$
\begin{aligned}
\text{Minimize} \quad & F(x,u) \\
\text{subject to} \quad & x = K(x,u) \\
& G(x,u) = 0 \\
& u \in U_{ad} = \{u \in L_\infty(0,T) : u_a(t) \le u(t) \le u_b(t)\},
\end{aligned}
\tag{2}
$$

with the operators $K$ and $G$ defined as follows:

$$
K \ : \ \Big(x(t), u(t)\Big) \mapsto c(t) + \int_o^t k(t,s) b(x(s), u(s), s)\, ds,
$$

$$
G \ : \ \Big(x(t), u(t)\Big) \mapsto \left( \int_0^T g_i(x(s), u(s), s)\, ds - c_i \right)_{i=1}^k,
$$

whereby

$$
K : X \times U \to X \qquad \text{and} \qquad G : X \times U \to R^k.
$$

The spaces involved hereto are defined as follows, $X := C[0,T]$ and $U := L_\infty(0,T)$, hence $X$ and $U$ are real Banach spaces. These operators, thanks to our assumptions, possess continuous first and second order Fréchet derivatives. We denote with $L(X,Y)$ the space of all linear continuous operators from $X$ into $Y$, and $L(X,X)$ will be abbreviated as $L(X)$. If $A \in L(X,Y)$ then $A^* \in L(Y^*, X^*)$ is the corresponding adjoint operator, with $Y^*$ and $X^*$ being the adjoint spaces of $Y$ and $X$ respectively. We denote with $K_x, G_x$ and $K_u, G_u$ the corresponding partial Fréchet derivatives of the operators $K, G$ at the point $(x_0, u_0)$ with respect to $x$ and $u$ respectively.

**Definition 2.1 (Admissible set)** *: The set*

$$
M = \{(x,u) \in C[0,T] \times L_\infty(0,T) : x = K(x,u), G(x,u) = 0, u \in U_{ad}\}
$$

*is called the admissible set to our problem (1) and (2).*

**Definition 2.2** *: $(x_0, u_0)$ is called local solution of the optimal control problem (2), if it is an admissible point, i.e. if $(x_0, u_0) \in M$ and $\exists \rho > 0$ such that for all $(x,u) \in M \cap \{(x_0, u_0) + B_\rho\}$ the following inequality is satisfied*

$$
F(x,u) \ge F(x_0, u_0)
\tag{3}
$$

*$B_\rho$ is the closed ball in $X \times U$ with radius $\rho > 0$ and centre at the origin.*

## 2.2 Regularity conditions and Lagrange multiplier rule

The Lagrange functional plays an important role in the formulation of the optimality conditions. With regard to our problem (2), the corresponding Lagrange functional is defined as follows:

$$L(x, u, y, z) = F(x, u) + <y, x - K(x, u)> + z^T G(x, u),$$

with $y \in C[0, T]^*$ and $z = (z_1, \cdots , z_k)^T \in R^k$ known as the Lagrange multipliers. The space $C[0, T]^*$ is the space of functions of bounded variations [BV] (cf: [17], p 135).

The regularity conditions, required for the existence of the Lagrange multiplier rule (and the Lagrange multipliers)are given as follows:

[R1] : $(x_0, u_0) \in C[0, T] \times L_\infty$ is said to be regular with regards to problem 2 if the following conditions are fulfilled:

$$\begin{aligned}
a) \quad & (I - K_x)^{-1} \in L(X, X) \quad \text{exists}, \quad (X = C[0, T]), \\
b) \quad & \left(G_x(I - K_x)^{-1}K_u + G_u\right) = 0 \quad \text{with}, \quad L_\infty(0, T) = R^k, \\
& \text{that is}, \quad G_x(I - K_x)^{-1}K_u + G_u \quad \text{is surjective}, \\
c) \quad & \exists x \in C[0, T], \quad \exists u \in U_{ad} : \\
& x = K_x(x_0, u_0)x + K_u(x_0, u_0)u, \\
& G_x(x_0, u_0)x + G_u(x_0, u_0)u = 0.
\end{aligned}$$

[R1] are regularity conditions for the optimal control problem 2. In what follows, we shall write, for short, $L_\infty^* := L_\infty(0, T)^*$ and $L_\infty := L_\infty(0, T)$.

**Lemma 2.1** *Let the regularity condition* [R1] *be fulfilled. Then there exist Lagrange multipliers* $(y, z) \in C[0, T]^* \times R^k$, *such that*

$$y = K_x^* y - F_x - G_x z \tag{4}$$

$$\langle F_u - K_u^* y + G_u z, u - u_0 \rangle_{L_\infty^*, L_\infty} \geq 0 \quad \forall u \in U_{ad} \tag{5}$$

(cf: [23], Theorem 1.3.1).

For the proof of this Lemma see [23], Lemma 1.3.1 and Theorem 1.3.1.

Lemma 2.1 above is also known as the multiplier rule and equation 4 is called the adjoint equation, while equation 5 is known as the variational inequality. The element $y$ is said to be the adjoint state.

## 2.3 Necessary and sufficient optimality conditions for the optimal control problem

The formulation of the first order necessary optimality conditions, as seen in the multiplier rule, leads us to Lagrange Multipliers and operators in complicated dual spaces, for instance, to the space of functions of bounded variations $C[0,T]^*$ and to $L_\infty(0,T)^*$ space. In order to avoid these complicated dual spaces we will embed them densely into bigger spaces, that possess simple dual spaces.

**Definition 2.3** *Let $X$ be a subset of another Banach space $\tilde{X}$. $X$ is said to be dense in $\tilde{X}$, if every element of $\tilde{X}$ can be represented as a limit of a convergent sequence of elements of $X$ (convergence in the sense of $\|\cdot\|_{\tilde{X}} - Norm$).*
**Continuous embedding:**
*Let $X,\tilde{X}$ be Banach spaces. If $X \subset \tilde{X}$ densely, and for a suitable number $c \in R$ the following relationship*

$$\|x\| \leq c\|x\|_X, \quad x \in X \tag{6}$$

*is fulfilled, then $X$ is said to be continuously embedded in $,\tilde{X}$.*

The operator $\Theta_x(X \to \tilde{X}) \in L(X,\tilde{X})$, defined by $\Theta_x(x) = x, \quad x \in X$ is called the embedding operator. Whence every element or function $x \in X$ is also an element of $\tilde{X}$. This operator $\Theta_x(x) = x$ with $x \in X$, maps every element $x$ to itself, whereby the resulting $x$ is understood to be an element of $\tilde{X}$. Therefore, $\Theta_x$ is an identity or unit operator. This operator is linear and bounded thanks to equation (6).
To formulate the first order necessary optimality conditions of our problem (1) and (2) respectively, we shall proceed along the lines of Tröltzsch [23] and Maurer [19] as follows: Considering the fact that the operators of our optimal control problem are twice partially differentiable in $U = L_\infty(0,T)$ and $X = C[0,T]$ only, with respect to their variables, we shall therefore carry out the differentiation as well as the formulation of the regularity conditions [**R1**] in $U$ and $X$. The resulting complicated dual spaces $X^*$ and $U^*$ can be avoided by embedding these spaces $X$ and $U$ densely and continuously in larger spaces $\tilde{X}$ and $\tilde{U}$ that possess simple dual spaces. We define these spaces $\tilde{X},\tilde{U}$ to be $L_p(0,T)$ spaces, with $(1 < p < \infty)$, and are going to be understood or written in a short form henceforth as $L_p$ spaces. In other words we extend $C[0,T]$ and $L_\infty(0,T)$ to $L_p(0,T)$, with $1 < p < \infty$. In this case we shall be dealing with a simple dual spaces such as $L_p(0,T)^* = L_q(0,T)$.
The space $L_q(0,T)$ is known as the dual space to $L_p(0,T)$, if $q = \frac{p}{p-1}$ or better still if $\frac{1}{p} + \frac{1}{q} = 1$ for $(1 < p < \infty)$. For example, the space of $L_2(0,T)$ is identcal to its dual space, that is $L_2(0,T) = L_2^*(0,T)$.
This idea of extending and embedding the operators and spaces is an important aspect in the verification of the optimality conditions. Let us once more take

a look at the spaces and operators involved:

$$F : C[0,T] \times L_\infty(0,T) \to R^1$$
$$K : C[0,T] \times L_\infty(0,T) \to C[0,T]$$
$$G : C[0,T] \times L_\infty(0,T) \to R^k.$$

REMARK: This embedding will be used only for the derivatives of our given quantities: We perform the differentiation in $U$ and $X$, afterwards the further derivatives are extended to operators in the $L_p$ spaces.

Representing the multiplier rule, we then deal with adjoint operators which are defined as follows:

$$K_x : C[0,T] \to C[0,T], \qquad K_x^* : C[0,T]^* \to C[0,T]^*,$$
$$K_u : L_\infty(0,T) \to C[0,T] \qquad K_u^* : C[0,T]^* \to L_\infty(0,T)^*,$$
$$G_x : C[0,T] \to R^k \qquad G_x^* : R^k \to C[0,T]^*,$$
$$G_u : L_\infty(0,T) \to R^k \qquad G_u^* : R^k \to L_\infty(0,T)^*,$$

whereby these dual spaces are difficult to represent. Instead of dealing with the $C[0,T]$ or $L_\infty(0,T)$ spaces we apply the $L_p(0,T)$ space ($1 \le p < \infty$), due to the fact that $C[0,T]$ and $L_\infty(0,T)$ are densely and continuously embedded in $L_p(0,T)$, that is

$$\exists c_1 > 0 : \|f\|_{L_p(0,T)} \le c_1 \|f\|_{L_\infty(0,T)} \qquad \forall f \in L_\infty(0,T). \tag{7}$$

In what follows, we present a precise definition of the first partial derivatives of the operators involved in our problem:

$$K_x : x(t) \mapsto \int_0^t k(t,s) b_x(x_0,u_0,s) x(s)\,ds \qquad (K_x \in L(C[0,T]))$$

$$K_u : u(t) \mapsto \int_0^t k(t,s) b_u(x_0,u_0,s) u(s)\,ds \qquad (K_u \in L(L_\infty(0,T), C[0,T]))$$

$$G_x : x(t) \mapsto \left( \int_0^T g_{i,x}(x_0,u_0,s) x(s)\,ds \right)_{i=1}^k \qquad (G_x \in L(C[0,T], R^k))$$

$$G_u : u(t) \mapsto \left( \int_0^T g_{i,u}(x_0,u_0,s) u(s)\,ds \right)_{i=1}^k \qquad (G_u \in L(L_\infty(0,T), R^k)).$$

We extend the operators $K_x : C[0,T] \to C[0,T], K_u : L_\infty \to C[0,T], G_x : C[0,T] \to R^k, G_u : L_\infty \to R^k$ continuously to $\check{K}_x, \check{K}_u : L_p \to C[0,T], \check{G}_u, \check{G}_x : L_p \to R^k$ respectively. Basically the operators $K_x, K_u$ and $\check{K}_x, \check{K}_u$ respectively, and $G_x, G_u$ and $\check{G}_x, \check{G}_u$ respectively are having the same form with a slight difference being in their domains. We can now remark that $\check{K}_x, \check{K}_u$ both map bigger spaces $L_p(0,T)$ to smaller ones $C[0,T]$. This property in function spaces is known as the smoothing property and holds in our case for $p \geq 1/(1-\alpha), \alpha \in (0,1)$. One could also consider the operator $K_x$ as one that maps $L_p(0,T)$ to $C[0,T]$, which in another way, its adjoint operator $K_x^*$ can be seen as one that maps $C[0,T]^*$ to a better dual space $L_p(0,T)^*$. We can still elaborate on this property by introducing the embedding operators $\Theta_x : C[0,T] \to L_p(0,T)$ and $\Theta_u : L_\infty(0,T) \to L_p(0,T)$, then

$$K_x = \check{K}_x \Theta_x, \qquad K_u = \check{K}_u \Theta_u.$$

Furthermore, the ranges of $\check{K}_x$ and $\check{K}_u$ are embedded in $L_p(0,T)$. Whence we define the operators $\hat{K}_x, \hat{K}_u : L_p \to Ł_p(0,T)$. The operators defined here, $K_x$ and $K_u$ as well as $K$ defined above, are Volterra integral operators. By the theory of Volterra integral equations, $(I - K_x)^{-1} \in L(C[0,T] \times C[0,T])$ does exist (see Gohberg & Goldberg [?], p. 230). The existence $(I - K_x)^{-1} \in L(L_p \times L_p)$ follows from the fact that the operator $K_x$ can be continuously extended to $\hat{K}_x$ (see [15]).

After introducing the smoothing property, we can now formulate and prove the first order optimality condition in a more convenient form.

**Theorem 2.1** *Let $(x_0, u_0)$ be a local optimal solution to our problem (2). Further, let $(x_0, u_0)$ be regular, i.e. the regularity condition* **[R1]** *is satisfied in $L_\infty(0,T)$ and $C[0,T]$. We also assume the continuous extension of the operators $K_x \in L(C[0,T])$ and $K_u \in L(L_\infty(0,T), C[0,T])$ to $\check{K}_x, \check{K}_u \in L(L_p(0,T), C[0,T])$ and furthermore to $\hat{K}_x, \hat{K}_u \in L(L_p(0,T))$ for $p > 1/(1-\alpha)$. Then there exist Lagrange multipliers $y \in L_p^*(0,T)$, $z \in R^k$, such that*

$$y = \hat{K}_x^* y - \check{K}_x^*(F_x + G_x^* z) \tag{8}$$
$$\langle -\hat{K}_u^* y + \check{K}_u^*(F_x + G_x^* z), u - u_0 \rangle_{L_q, L_p} + \langle \hat{F}_u + G_u^* z, u - u_0 \rangle_{X^*, X} \geq 0$$
$$\forall u \in U_{ad} \tag{9}$$

*are satisfied.*

The results of Theorem 2.1 could also be obtained formally through the Lagrange functional under the condition that the regularity conditions **[R1]** are satisfied. one presents the Lagrangian, with respect to (1), in integral form as follows:

$$\begin{aligned}
L(x,u,y,z) &= \int_0^T f(x(t), u(t), t) dt \\
&= \int_0^T y(t)\Big(x(t) - c(t) - \int_0^t k(t,s) b(x(s), u(s), s) ds\Big) dt \\
&\quad + \sum_{i=1}^k z_i \Big(\int_0^T g_i(x(t), u(t), t) dt - c_i\Big),
\end{aligned}$$

with $\quad z = (z_1, \cdots, z_k)^T, \qquad (y, z) \in L_p^*(0, T) \times R^k.$

The adjoint equation is obtained from

$$L_x(x_0, u_0, y, z)h = \int_0^T f_x(x_0(t), u_0(t), t)h(t)dt \tag{10}$$

$$+ \int_0^T y(t)\Big(h(t) - \int_0^t k(t, s)b_x(x_0(s), u_0(s), s)h(s)ds\Big)dt$$

$$+ \int_0^T \sum_{i=1}^k z_i g_{i,x}(x_0(t), u_0(t), t)h(t)dt = 0 \qquad \forall h \in C[0, T].$$

Whence,

$$y(t) = b_x(x_0, u_0, t)\int_t^T k(t, s)y(s)ds - f_x(x_0, u_0, t) - \sum_{i=1}^k z_i g_{i,x}(x_0, u_0, t). \tag{11}$$

Equation (11) is known as the adjoint equation to our control problem (1). And the variational inequality $L_u(x_0, u_0, y, z)(u - u_0) \geq 0$ obtains the form

$$L_u(x_0, u_0, y, z)(u - u_0) = \int_0^T f_u(x_0(t), u_0(t), t)(u - u_0)(t)dt$$

$$+ \int_0^T y(t)\Big(-\int_0^t k(t, s)b_u(x_0(s), u_0(s), s)(u - u_0)(s)ds\Big)dt$$

$$+ \int_0^T \sum_{i=1}^k z_i g_{i,u}(x_0(t), u_0(t), t)(u - u_0)(t)dt.$$

Whence,

$$\int_0^T \Big(f_u(.,.) + \sum_{i=1}^k z_i g_{i,u}(.,.) - b_u(.,.)\int_t^T k(t, s)y(s)ds\Big)(u - u_0)(t)dt \geq 0. \tag{12}$$

This adjoint equation (11), when compared with the case of Pontryagin maximum principle for the case of control of ordinary differential equations (we obtain it by setting $k(t, s) = 1$), is not the standard type.

Therefore we introduce a new adjoint state $y_0(t)$ and define it as follows:

$$y_0(t) = \int_t^T k(s, t)y(s)ds \tag{13}$$

The adjoint equation (11) now reads

$$y(t) = b_x(x_0, u_0, t)y_0(t) - f_x(x_0, u_0, t) - \sum_{i=1}^k z_i g_{i,x}(x_0, u_0, t). \tag{14}$$

Substituting $y(t)$ of equation (14) in (13) delivers the following new adjoint equation:

$$y_0(t) = \int_t^T k(t,s)\Big(b_x(x_0,u_0,s)y_0(s)$$

$$-f_x(x_0(s),u_0(s),s) - \sum_{i=1}^k z_i g_{i,x}(x_0(s),u_0(s),s)\Big)ds. \qquad (15)$$

The variational inequality then reads

$$\int_0^T \Big(f_u(t) + \sum_{i=1}^k z_i g_{i,u}(t) - b_u(t)y_0(t)\Big)(u-u_0)(t)dt \geq 0, \qquad (16)$$

where
$$b_u(t) := b_u(x_0(t),u_0(t),t), \quad f_u := f_u(x_0(t),u_0(t),t), \text{ and } g_{i,u} := g_{i,u}(x_0(t),u_0(t),t).$$

# 3 The conditioned gradient method for linear-quadratic control problems

We devote this section to the implementation of the conditioned gradient method to the solution of an optimal control problem of linear quadratic type, making use of the results in the previous section as well as the numerical method for solving weakly singular Volterra integral equations handled in chapter 2. In what follows, we consider a special class of optimal control problems:

$$(P) \qquad \text{Minimize} \quad \int_0^T f(x(t),u(t),t)$$

$$\text{subject to} \quad x(t) = c(t) + \int_0^t (t-s)^{-\alpha}[x(s)+u(s)]ds \qquad (17)$$

$$u \in U_{ad}, \quad \text{with} \quad u_a(t) = -1, \quad u_b = +1$$

Let $(x_0,u_0)$ be the optimal solution of problem (17). Applying the theory of Section 3.4 we construct the Lagrangian as follows

$$L(x,u;y) = \int_0^T f(x(t),u(t),t)dt \qquad (18)$$

$$- \int_0^T y(t)\Big\{x(t) - c(t) - \int_0^t (t-s)^{-\alpha}[x(s)+u(s)]ds\Big\}dt,$$

and obtain, at $(x_0,u_0)$, after application of the chain rule and Fubini's theorem, the following adjoint equation

$$y(t) = f_x(x_0(t),u_0(t),t)dt + \int_t^T (s-t)^{-\alpha}y(s)ds, \qquad (19)$$

which could be transformed to equation (15). Furthermore we obtain the variational inequality

$$L_u(x_0, u_0; y)(u - \bar{u}) = \int_0^T \left( f_u(x_0, u_0, t) + \int_t^T (s - t)^{-\alpha} y(s) ds \right)(u - \bar{u}) dt, \quad (20)$$

with

$$L_u(x_0, u_0; y)(u - \bar{u}) \geq 0 \qquad \forall u \in U_{ad}.$$

We now consider the condition gradient method algorithm for solving the above problem:

**Gradient - Method - Algorithm:**
**Step 1** Set $k = 0$, choose a start solution $u_k$.
**Step 2** compute $x_k$ as a solution of the state equation

$$x(t) = c(t) + \int_0^t (t - s)^{-\alpha} [x(s) + u(s)] ds$$

with $u_k$ and the computed $x_k$ determine the value

$$F(x_k, u_k) = \int_0^T f(x_k(t), u_k(t), t) dt.$$

**Step 3** compute $y_k$ as a solution of the adjoint equation (19) by backward integration, making use of the acquired $x_k$ in step 2.
**Step 4** Determine the direction of descent as follows: If $u_k$ would be optimal, then

$$\left( L_u(x_k, u_k; y_k), (u - u_k) \right) \geq o \quad \forall u \in U_{ad}$$

$$\Leftrightarrow \int_0^T L_u(x_k, u_k; y_k)(u - u_k)(t) dt \geq 0 \quad \forall u \in [u_a(t), u_b(t)] \text{ and } a.a. \ t \in [0, T]$$

$$\Leftrightarrow \min_{u \in U_{ad}} \int_0^T L_u(x_k, u_k; y_k) u(t) dt = \int_0^T L_u(x_k, u_k; y_k) u_k(t) dt$$

$$\Leftrightarrow \min_{-1 \leq u \leq 1} L_u(x_k, u_k; y_k) u(t) = L_u(x_k, u_k; y_k) u_k(t) \text{ for a.a } t \in [0, T].$$

This then provides the idea of finding a new direction of descent, $v(t)$, if $u_k$ is not optimal. We then determine $v(t)$ by

$$\min_{-1 \leq u \leq 1} L_u(x_k, u_k; y_k) u(t) = L_u(x_k, u_k; y_k) v(t) \text{ for almost all } t \in [0, T],$$

and compute $L_u(x_k, u_k; y_k)(t)$. Hence

$$L_u(x_k, u_k; y_k)(t) \begin{cases} < 0 & \Rightarrow v(t) = 1, \\ = 0 & \Rightarrow \text{ we set } v(t) = 0, \\ > 0 & \Rightarrow v(t) = -1. \end{cases}$$

$v(t)$ gives a direction of descent.

**Step 5** Determine the step size as follows: we minimize the cost function with respect to $\gamma$, whereby

$$u_{k+1}(t) = (1 - \gamma)u_k(t) + \gamma v(t), \qquad [0, 1].$$

With the new control $u_{k+1}$ we compute the new state $x_{k+1}$ and evaluate the cost function $F(x_{k+1}, u_{k+1})$. $\gamma$ delivers the step size in the direction of descent. The initial $\gamma$ would be given and the subsequent ones will be determined as indicated in **Step 6**.

**Step 6**: If $F(x_k, u_k) \leq F(x_{k+1}, u_{k+1}) \quad \Rightarrow$ then set $\gamma = \frac{\gamma}{2}$ Go to **Step 5**.
$F(x_k, u_k) > F(x_{k+1}, u_{k+1})$ then set $k := k + 1$ and go to **Step 2**.
$F(x_k, u_k) = F(x_{k+1}, u_{k+1})$ or $\|F(x_k, u_k) - F(x_{k+1}, u_{k+1})\| \leq 10^{-6}$ then stop.

**Remark**:
Notice that $u_{k+1}$ is a convex combination of $u_k \in U_{ad}$ and $v \in U_{ad}$, hence $u_{k+1}$ is also an element of $U_{ad}$. This is an advantage of the conditioned gradient method.

## 3.1 Dicretization and numerical implementation

The state equations (17) and the adjoint equation (19) both describe Volterra integral equations. Thence for their computation we will apply the collocation method described in section 2.3 of chapter 2. The time interval $[0, T]$ will be discretized into a finite number of subintervals as indicated in Figure 1.

$$[0, T] = \bigcup_{j=0}^{Noint-1} [T_j, T_{j+1}),$$

with $Noint = $ number of (equidistant) subintervals. $Tj$ is obtained as follows

$$T_j = \frac{T}{Noint} * j, \qquad \text{for} \quad j = 0, \cdots, Noint.$$

The control function, $u(t)$, describes a step function and it is constant in each subinterval. Thus $u_{j+1} \in [T_j, T_{j+1})$. This discretization is in conformity with the conditions of Theorem 2.3 and Theorem 2.4, which state that the convergence of the method applied can be guaranteed only when the functions under the integral signs are continuous in the required interval. In order to abide by this convergence rule we consider the subintervals as "independent" and separate entities, and in the same way we compute the state equation (17) and the adjoint equation (19) in these intervals. Each subinterval $[T_j, T_{j+1})$ would be discretized further along the lines of section 2.3 to obtain the discrete points

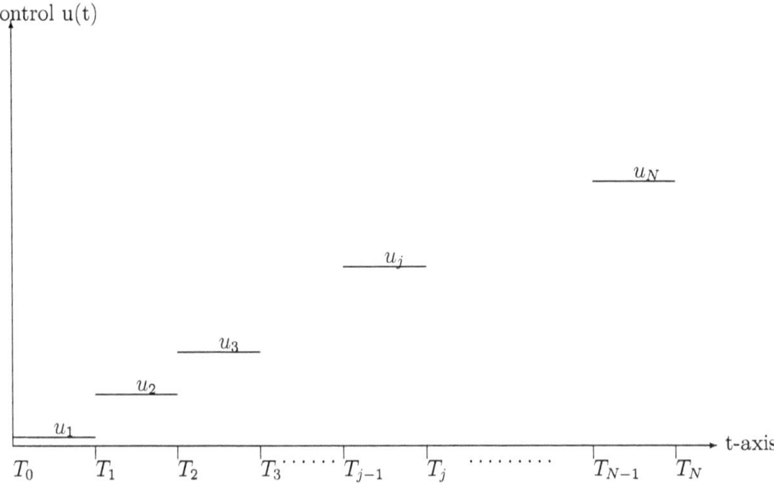


Figure 1: Graphical representation of the step (control) functions

of $t$. The discrete points $\{t_k\}_{k=0}^N$ of an independent subinterval in the forward integration (17) would read

$$t_k = (T_{j+1} - T_j) * \left(\frac{k}{N}\right)^r + T_j,$$

while those in the backwards integration would be given as follows:

$$t_k = \frac{T}{Noint} - (T_{j+1} - T_j) * \left(\frac{k}{N}\right)^r + T_j, \qquad .$$

Noint = number of subintervals in $[0, T]$, $r \geq 1$, for instance $r = \frac{2}{1-\alpha}$, see Section 2.3.3, and $j = 0, \cdots, Noint - 1$.

## 3.2   Computation of the state equation

We rewrite our state equation

$$x(t) = c(t) + \int_0^t (t - s)^{-\alpha}[x(s) + u(s)]ds. \tag{21}$$

Its discretization is almost identical to the one described in Section 2.3.2, but with a slight difference. In Section 2.3.2 the time interval $[0, T]$ was discretized as an entity, that is, without sub-divisions, whereas here we shall be dealing with sub-intervals of $[0, T]$. Here the control function is given to be constant (continuous) in each sub-interval. We proceed, in this regard as follows:
**First Interval** $t \in [0, T_1]$, here the defined control $u_1$ is identically constant:

$$x(t) = c(t) + \int_0^t (t - s)^{-\alpha}[x(s) + u_1(s)]ds.$$

13

**Second Interval** $t \in [T_1, T_2]$, with $u_2$ being the control (identically constant):

$$x(t) = \underbrace{c(t) + \int_0^{T_1} (t-s)^{-\alpha}[x(s) + u_1(s)]ds}_{\tilde{c}(t)} + \int_{T_1}^t (t-s)^{-\alpha}[x(s) + u_2(s)]ds.$$

Hence

$$x(t) = \tilde{c}(t) + \int_{T_1}^t (t-s)^{-\alpha}[x(s) + u_2(s)]ds.$$

**Third Interval** $t \in [T_2, T_3]$, with $u_3$ being the control:

$$x(t) = \underbrace{c(t) + \int_0^{T_1} (t-s)^{-\alpha}[x(s) + u_1(s)]ds + \int_{T_1}^{T_2} (t-s)^{-\alpha}[x(s) + u_2(s)]ds}_{\tilde{c}(t)}$$
$$+ \int_{T_2}^t (t-s)^{-\alpha}[x(s) + u_3(s)]ds.$$

Whence

$$x(t) = \tilde{c}(t) + \int_{T_2}^t (t-s)^{-\alpha}[x(s) + u_3(s)]ds,$$

and so on till the interval $t \in [T_{Noint-2}, T_{Noint-1}]$ is reached.
We now derive a general equation for all these stages on $[T_0, T_{Noint}]$:

$$x(t) = \underbrace{c(t) + \int_0^{T_1} (..)^{-\alpha}[x(s) + u_1(s)]ds + \int_{T_1}^{T_2} (..)^{-\alpha}[x(s) + u_2(s)]ds + \cdots +}_{\tilde{c}(t)}$$
$$+ \int_{T_j}^t (t-s)^{-\alpha}[x(s) + u_{j+1}(s)]ds. \tag{22}$$

The general formula for he discretized $\tilde{c}(t_{kj})$ reads:

$$\tilde{c}(t_{kj}) = \sum_{z=0}^{j-1} \frac{1}{1-\alpha} \sum_{i=0}^{N-1} \left[ x_{i+N*z} + u_{i+N*z} \right]\left[ (t_{kj} - s_i^z)^{1-\alpha} - (t_{kj} - s_{i+1}^z)^{1-\alpha} \right], \tag{23}$$

whereby these $S_i^z$ represent the discrete points in the preceeding subintervals.
We apply the method illustrated in Section 2.3 to compute equation(23). From
this equation we obtain $x$, analoguous to Section @.3, from the Vector-Matrix-
Equation

$$\underbrace{\left( I - h_k^{1-\alpha} W \right)}_{Q} x = d. \tag{24}$$

Whereby

$$d = \left( \tilde{c}(t_{k1} + \sum_{i=0}^{k-1} h_i^{1-\alpha}(\cdots), \quad \tilde{c}(t_{k2} + \sum_{i=0}^{k-1} h_i^{1-\alpha}(\cdots) \right)^T ;$$

with

$$x = Q^{-1}d, \quad \text{provided} \quad Q_{-1} \text{ does exist.}$$

## 3.3 Computation of the adjoint equation

After partitioning of the interval $[0, T]$ into subintervals, the adjoint equation (19) can be written as follows:

$$
\begin{aligned}
y(t) \;=\;& f_x(x(t), u(t), t) + \int_{T_{Noint-1}}^{T} (s-t)^{-\alpha} y(s) ds \\
& \int_{T_{Noint-2}}^{T_{Noint-1}} (s-t)^{-\alpha} y(s) ds + \cdots + \int_{T_{Noint-j-1}}^{T_{Noint-j}} (s-t)^{-\alpha} y(s) ds \\
& + \cdots + \int_{T_1}^{T_2} (s-t)^{-\alpha} y(s) ds + \int_{t}^{T_1} (s-t)^{-\alpha} y(s) ds.
\end{aligned}
\tag{25}
$$

We set

$$
\begin{aligned}
\hat{f}(t) \;=\;& f_x(x(t), u(t), t) + \int_{T_{Noint-1}}^{T} (s-t)^{-\alpha} y(s) ds \\
& \int_{T_{Noint-2}}^{T_{Noint-1}} (s-t)^{-\alpha} y(s) ds + \cdots + \int_{T_{Noint-j-1}}^{T_{Noint-j}} (s-t)^{-\alpha} y(s) ds,
\end{aligned}
$$

and obtain a general equation of the form

$$y(t) = \hat{f}(t) + \int_{t}^{T_j} (s-t)^{-\alpha} y(s) ds \qquad j = Noint, \cdots, 0. \tag{26}$$

Contrary to the computation of the state equation, the adjoint equations will be computed backwards. It must also be remarked that the collocation points in the forward integration process are different from those in the backward process in the case of asymmetric graded meshes, despite the fact that they both carry, for simplicity, the same notation $t_k$. In this case we have to interpolate the result of the state equation so as to obtain the discrete values of $x(t)$ at the corresponding collocation points for the backward integration. The Volterra equation (26) would be computed following the proceedings of Section 2.3, but setting $t = t_{kj} = t_k - c_j h_k$, with $k = 0, \cdots, N-1$ and $h_k = t_k - t_{k+1}$. From (26) we obtain

$$y(t_{kj}) = \hat{f}(t_{kj}) + \int_{t_{kj}}^{t_k} (t-s)^{-\alpha} y_k(s) ds + \sum_{i=0}^{k-1} \int_{t_{i+1}}^{t_i} (s-t)^{-\alpha} y_i(s) ds,$$

for $k = 0, \cdots, N-1$, (take note that $t_0 = T$, and $t_N = 0$).
After substituting $s := t_i - vh_i$ and changing the variable under the integration sign, we obtain, in a compact form

$$y(t_{kj}) = \hat{f}(t_{kj}) + h_k^{1-\alpha}\Psi_{kk} + \sum_{i=0}^{k-1} h_i^{1-\alpha}\Psi_{ki}. \tag{27}$$

Hereby is

$$\Psi_{ki} = \sum_{l=1}^{m} w_{jl}^{ki} y_{il} \qquad i = 0, \cdots, N-1,$$

$$\Psi_{kk} = \sum_{l=1}^{m} w_{jl} \sum_{s=1}^{2} L_s(c_j c_l) y_{ks} \qquad i = 0, \cdots, N-1,$$

$$w_{jl}^{ki} = \int_0^1 \left(\frac{t_i - t_{kj}}{h} - v\right)^{-\alpha} L_l(v)\,dv; \quad \text{with } L_l(v) = \prod_{k=1}^{m} \frac{v - c_k}{c_l - c_k},$$

$$w_{jl} = \int_0^{c_j} (c_j - v)^{\alpha} L_{jl}(v)\,dv; \quad \text{with } L_{jl}(v) = \prod_{k=1}^{m} \frac{v - c_j c_k}{c_j c_l - c_j c_k}, \quad k \neq l.$$

For a specific m, for instance $m = 2$ (see Section 2.3) we calculate $w_{jl}^{ki}$ and $w_{jl}$ analytically for $l = 1, 2$ and obtain the following:

$$w_{j1} = \frac{c_j^{1-\alpha}[1 - c_2(2-\alpha)]}{(c_1 - c_2)(1-\alpha)(2-\alpha)}; \qquad w_{j2} = \frac{c_j^{1-\alpha}[1 - c_1(2-\alpha)]}{(c_2 - c_1)(1-\alpha)(2-\alpha)};$$

$$w_{j1}^{ki} = \frac{1}{(c_1 - c_2)\eta}\left[(c_2 - 1)(2-\alpha)\theta^{1-\alpha} - \theta^{2-\alpha} - (1-\alpha)c_2\phi^{1-\alpha} + \phi^{2-\alpha}\right];$$

$$w_{j2}^{ki} = \frac{1}{(c_2 - c_1)\eta}\left[(c_1 - 1)(2-\alpha)\theta^{1-\alpha} - \theta^{2-\alpha} - (2-\alpha)c_1\phi^{1-\alpha} + \phi^{2-\alpha}\right];$$

$$\phi = \frac{t_i - t_{kj}}{h_i}; \qquad \theta = \left(\frac{t_i - t_{kj}}{h_i} - 1\right); \qquad \eta = (1-\alpha)(2-\alpha).$$

Equation (27) now reads, for $(j = 1, 2)$

$$\hat{y}_{kj} - h_k^{1-\alpha}\left(w_{j1}\left[L_1(c_j c_1)\hat{y}_{k1} + L_2(c_j c_1)\hat{y}_{k2}\right] + w_{j2}\left[L_1(c_j c_2)\hat{y}_{k1} + L_2(c_j c_2)\hat{y}_{k2}\right]\right)$$

$$= \hat{f}(t_{kj}) + \sum_{i=0}^{k-1} h_i^{1-\alpha}\left(w_{j1}^{ki}\hat{y}_{i1} + w_{j2}^{ki}\hat{y}_{i2}\right). \tag{28}$$

Hereby

$$\hat{f}(t_{kj}) = f_x(t_{kj}) + \frac{1}{1-\alpha} \sum_{z=j+1}^{Noint-1} \sum_{i=0}^{N-1} y_{i+N*(Noint-1-z)}$$

$$\cdot \left[(s_z^i - t_{kj})^{1-\alpha} - (s_z^{i+1} - t_{kj})^{1-\alpha}\right], \tag{29}$$

with $s_z^i$ being the collocation points at the preceding intervals.
Equation (28) in vector-matrix-form can be written as

$$\underbrace{I - h_k^{1-\alpha}Q}_{Z}\, y = b. \tag{30}$$

Hence we can write

$$y = Z^{-1}b, \ \ \text{provided } Z^{-1} \text{ does exist, with } y = (\hat{y}_{k1}, \hat{y}_{k2}).$$

The values of $y(t)$ will be approximated in the intervals $\sigma_k = [t_{k+1}, t_k]$ as follows:

$$y(t_k - vh_k) = \sum_{j-1}^{2} L_j(v)\hat{y}_{kj}, \ \text{for } v \in [0, 1].$$

## 3.4 Determination of the direction of descent and computation of the cost function

The direction of descent is given as per STEP 4 of the algorithm in Section 3.5:

$$L_u(x_x, u_k; y_k)v = \int_0^T \left( f_u(x_k, u_k, t) + \int_t^T (s - t)^{-\alpha} y_k(s)ds \right)v(t)dt. \tag{31}$$

From the adjoint equation (19) we obtain

$$\int_t^T (s - t)^{-\alpha} y_k(s)ds = y(t) - f_x(x_k, u_k, t).$$

Setting this in (31)delivers

$$L_u(x_k, u_k; y_k)v = \int_0^T \left[ f_u(x_k, u_k, t) + y_k(t) - f_x(x_k, u_k, t) \right]v(t)dt.$$

We determine the value in braces [...] at the collocation points $t_k$, ($k = 0, \cdots, N - 1$) in the subintervals 0f $[0, T]$. This gives us the descent direction $v(t)$ at the corresponding mesh points, with

$$v(t_k) = \begin{cases} -1 & \text{if } [...] > 0, \\ 1 & \text{if } [...] < 0, \\ 0 & \text{if } [...] = 0. \end{cases}$$

The computation of the cost function

$$\int_0^T f(x(t), u(t), t)dt$$

in $[0, T]$ will be achieved by applying the trapezoidal rule. The computation formula in one of the subintervals $[T_j, T_{j+1}]$, with mesh points $t_k$ reads

$$\begin{aligned}
\kappa_j &= \int_{T_j}^{T_{j+1}} f(x(t), u(t), t)dt \\
&= \sum_{k=0}^{N-1} \frac{h_k}{2}\Big[f(x_{[k+N*j]}, u_j, t_k) + f(x_{[k+N*j+1]}, u_{j+1}, t_k)\Big],
\end{aligned}$$

with $h_k$ being the $k^{th}$ mesh length. We obtain as an approximation of the cost function over the entire interval $[0, T]$ the following:

$$\int_0^T f(x(t), u(t), t)dt = \sum_{j=0}^{Noint-1} \kappa_j.$$

## 4 Numerical example

In our numerical example we tested an optimal control problem of the following class:

$$\begin{aligned}
\text{Minimize} \quad & \int_0^T \big(x(t) - q(t)\big)^2 dt + \lambda \int_0^T u(t)^2 dt \\
\text{subject to} \quad & x(t) = c(t) + \int_0^t (t-s)^{-\alpha}[x(s) + u(s)]ds \qquad (32) \\
& -1 \le u(t) \le 1
\end{aligned}$$

The functions $c(t), q(t)$ as well as the values of $\alpha$ and $\lambda$ are given in this specific example. We attributed the following values and functions:

$$T = 1, \qquad q(t) = t; \qquad c(t) = t - \frac{4}{3}t^{\frac{3}{2}}.$$
$$\alpha = 0.5 \text{ and } \lambda = 10^{-1}.$$

Our numerical computation had the following settings and results:
number of subintervals $j$ in $[0, T] = 12$,
number of collocation points $N$ per subinterval $= 10$, hence a total of 120 collocation points on the entire $[0, T]$,
initial control function value $u(t) = 0.4$ constant in $[0, T]$,
initial stepsize $\gamma = 0.7$.
With this setting we obtain an exact solution of (32), whereby the cost functional is minimized to zero $(= 0)$, with a control function $u(t) = 0$ and a state function $x(t) = t$ in the entire domain $[0, T]$, see Section 2 Example 1a.

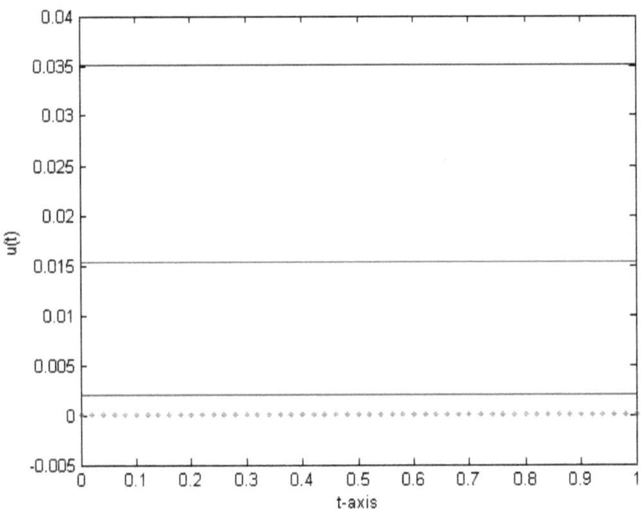

Figure 2: Sequence of the control function

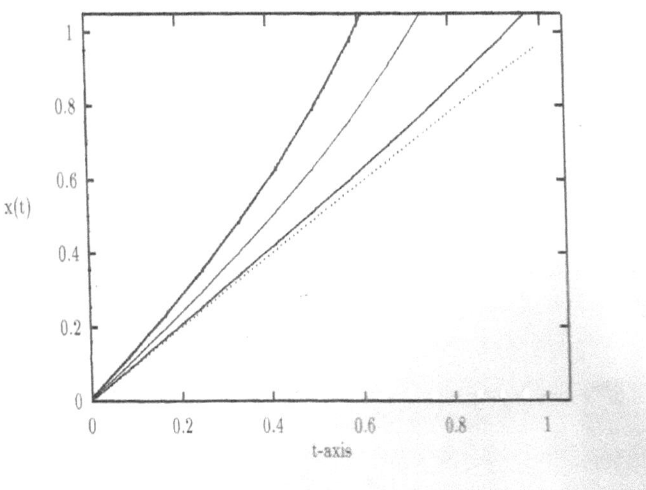

Figure 3: Sequence of the state function

**Remarks** After 8 iterations we obtained: minimum = 0.000233, see Figure 2 and Figure 3 for the representation of the control and state functions, respectively. Only the last four iteration steps have been represented graphically. The graph or function represented by dots is a portrait of the final solution of the control and/or state function, respectively, after 8 iterations.

# References


[1] Alt, W., 1992. *Sequential quadratic programmin in Banach spaces. In Oettli, W. and Pallaschke, D. (Eds), Advances in Optimization. Lecture notes in economics and mathematical systems Vol. 382, Springer Verlag, 281-301.*

[2] Alt, W., 1990. The Lagrange-Newton method for infinite-dimensional optimization problems. *Numer. Functional Analysis and Optimization*, 11:201-224.

[3] Alt, W. and Malanowski, K., 1993. The Lagrange-Newton method for nonlinear optimal control problems. *Computational Optimization and Applications* 2, pp 77-100.

[4] Alt, W., Sonntag, R. and Tröltzsch, F., (1996). An SQP method for optimal control of weakly singular Hammerstein integral equations. *Appl. Math. Optimization 33* (1996), 227-252.

[5] Brunner, H. (2004).*Collocation methods for Volterra integral and related functional differential equations*, Cambridge University Press, Cambridge.

[6] Brunner, H. (1987). Collocation methods for one-dimensional Fredholm and Volterra integral equations. *In The State of the Art in Numerical Analysis (A. Iserles and M.J.D. Powell, eds.)*, Clarendon Press, Oxford, pp. 563-600.

[7] Brunner, H. (1985). The numerical solution of weakly singular Volterra equations by collocation on graded meshes. *Mathematics of Computation* Vol. 45 No. 172, pp 417-437.

[8] Brunner, H. (1983). Nonpolynomial spline collocation for Volterra equations with weakly singular kernels. *SIAM J. Numerical Analysis* 20, 1106-1119.

[9] Brunner, H., Hairer, E. and Norsett, S. P. (1982). Runge-Kutta theory for Volterra integral equations of the second kind. *Mathematics of Computation* 39, 147-163

[10] Brunner, H., Pedas, A. and Vainikko, G. (1999). The piecewise polynomial collocation method for nonlinear weakly singular Volterra equations. *Math. Comp.*, 68, 1079 - 1095.

[11] Brunner, H. and van der Houwen, P. J. (1986). The numerical solution of Volterra equations. *CWI Monographs 3 (Centrum voor Wiskunde en Computer Science)* North Holland.

[12] Casas E., Trölztsch, F. and Unger, A., (1996). Second order sufficient optimality condition for a nonlinear elliptical control problem. *J. for Analysis and its Applications (ZAA) 15, pp 687-707.*

[13] de Hoog, F. and Weiss, R. (1973). On the solution of Volterra integral equation with a weakly singular kernel. *SIAM J. Math Anal.*, 4, 561-573.

[14] Diogo, T. and Lima, P. (2007). Collocation solution of weakly singular Volterra integral equation. *TEMA Tend. Mat. Apl. Comput.* 8 No. 2, 229-238.

[15] Goldberg and Tröltzsch, F., (1993)Second order sufficient optimality conditions for a class of non-linear parabolic boundary control problems. *SIAM J. Contr. Opt. 31 (1993)*, pp 1007-1025. the smooth solution of linear Volterra integral equations. *Int. J. Nonlinear Anal. Appl.* 4 No. 2, 1-10.

[16] Linz, P. (1985). Analytical and numerical methods for Volterra equations. *SIAM* Philadelphia

[17] Ljusternik, L. A., and Sobolew, W. I. (1976). *Elements of functional analysis.* Akademie Verlag, Berlin.

[18] Makroglou, A., (1981, July). A block-by-block method for Volterra integro-differential equations with weakly singular kernels. *Mathematics of Computation* Vol. 37 Number 155, pp 95-99.

[19] Maura, H., (1981). First and second order sufficient optimality conditions in mathematical programming and optimal control. *Mathematical Programming Study* 14, pp 163-177.

[20] Maurer, H., and Zowe, J. (1979). First and second order necessary and sufficient optimality conditions for infinite dimensional programming problems. *Mathematical Programming*, 16, pp 98-110.

[21] Schmidt, W. H., (1993) Iterative methods for optimal control processes governed by integral equations. *International Series of Numerical Mathematics*, 111 (1993) 6982.

[22] Te Riele, H. J. J., (1982). Collocation methods for weakly singular second kind Volterra integral equations with non-smooth solution. *IMA Jounal of Numerical Analysis* 2, pp. 437-449

[23] Tröltzsch, F., (1994) An SQP method for optimal control of a nonlinear heat equation. *Control and Cybernetics 23*, pp 267-288.

[24] Tröltzsch, F., (1985) On changing the spaces in Lagrange multiplier rules for the optimal control of nonlinear operator equations. *Optimization* 16, pp. 877-885.

[25] Tröltzsch, F., (1984) *Optimality conditions for parabolic control problems and applications.* Teubner-Texte zur Mathematik, vol. 62, Teubner Verlag, Leipzig.

[26] Tröltzsch, F. and Rösch, A., (2006) Existence of regular Lagrange multipliers for a nonlinear elliptic optimal control problem with pointwise control-state constraints. *SIAM J. Control and Optimization* 45, pp 548-564.

[27] Tröltzsch, F. and Rösch, A., (2003) Sufficient second order optimality condititions for a parabolic optimal control problem with pointwise control-state constraints. *SIAM J. Control and Optimization* 42, pp 138-154.

[28] Vainikko, G. (1976). *Funktionalanalysis der Diskretisierungsmethoden.* Teubner Verlag, Leipzig.

# YOUR KNOWLEDGE HAS VALUE

- We will publish your bachelor's and
  master's thesis, essays and papers

- Your own eBook and book -
  sold worldwide in all relevant shops

- Earn money with each sale

Upload your text at www.GRIN.com
and publish for free